THIS BOOK BELONGS TO:

_____

Dedicated to all the axolotl lovers.

All rights reserved.
No part of this book may be reproduced in any form or by any means, electronic or mechanical, and no photocopying or recording, unless you have written permission from the author.

ISBN 978-1-958985-29-8

Text copyright © 2024 by Mimi Jones

www.joeysavestheday.com

A Mimi Book

Axolotls, often called the Mexican Walking Fish, are not fish at all but a type of neotenic salamander.

They stay looking like babies their entire life.

These fascinating creatures have captured the imagination of scientists and animal lovers alike with their unique features and regenerative abilities.

They have a unique ability to regrow their lost body parts. When they lose a limb, they can regrow it in as little as three weeks. It's a process where nearby cells travel to the injury site, and regeneration starts.

Axolotls possess the remarkable ability to regenerate limbs and organs, including their hearts and sections of their brains, all without leaving any scars. This incredible characteristic has turned them into a subject of extensive scientific research.

They originate from Mexico's Lake Xochimilco and Lake Chalco. Axolotls inhabit freshwater lakes and ponds in Mexico.

Axolotls are unique among amphibians because they don't change into land-dwelling adults. They keep their young features, like gills and a tail, throughout their lives. This condition is known as neoteny.

In the wild, axolotls have a diverse diet, feeding on small fish, mollusks, arthropods, terrestrial worms, salmon eggs, and zooplankton.

The name "axolotl" comes from the Aztec language, Nahuatl, and is often associated with the Aztec god Xolotl, who could transform into an axolotl to escape his enemies.

Wild axolotls are usually brownish-gray, while those in captivity are often white with pink gills. They have feathery gills, which they keep along with fully functional lungs, allowing them to breathe underwater.

Although axolotls are endangered in their natural habitat, they have gained immense popularity as pets around the globe. In the pet trade, they are bred in a variety of colors, particularly in lighter hues.

Axolotls possess an upward-curving mouth, which gives them a perpetual smile. Their gentle disposition makes them a favorite among pet owners.

Conservationists are diligently striving to support axolotls by restoring their habitats and assisting in the breeding of individuals that can be reintroduced into the wild.

Axolotls have a lifespan that can extend up to 15 years.

Axolotls are carnivorous creatures that feed on worms, small fish, and a range of other aquatic delicacies.

Axolotls lack eyelids and therefore sleep with their eyes wide open.

Axolotls are cold-blooded creatures that depend on their environment to regulate their body temperature.

They thrive best in water temperatures between 60 and 68 degrees Fahrenheit.

Axolotls have feathery gills on their heads, which help them breathe underwater

Axolotls also have lungs that allow them to breathe by gulping air from the water's surface.

Female axolotls can lay up to one thousand eggs at once—talk about a big family! These eggs sink to the lake bottom and hatch within two weeks.

Axolotls can change their colors to blend in, avoid predators, or sneak up on prey. They share this clever trick with their salamander relatives.

Axolotls are nocturnal creatures, so they're more active during the night.

Scientific Classification of Axolotls

Scientific name: Ambystoma mexicanum

Kingdom:	Animalia
Phylum:	Chordata
Class:	Amphibia
Order:	Urodela
Family:	Ambystomatidae
Genus:	Ambystoma

The axolotl's extraordinary skill to regrow lost body parts has fascinated both scientists and the general public, presenting possibilities for progress in human tissue regeneration.

As we learn more from these extraordinary creatures, we must remember the significance of conserving their natural habitats to ensure their survival for future generations.

Count the Axolotls.

The end!

www.ingramcontent.com/pod-product-compliance
Lightning Source LLC
Chambersburg PA
CBHW040029050426
42453CB00002B/58